Pavel Astafiev

TRIZ and ARIZ

Pavel Astafiev

TRIZ and ARIZ

Basic Technological Techniques for Building Patent Protection for Smart Home Infrastructure Components and Structures

ScienciaScripts

Imprint
Any brand names and product names mentioned in this book are subject to trademark, brand or patent protection and are trademarks or registered trademarks of their respective holders. The use of brand names, product names, common names, trade names, product descriptions etc. even without a particular marking in this work is in no way to be construed to mean that such names may be regarded as unrestricted in respect of trademark and brand protection legislation and could thus be used by anyone.

Cover image: www.ingimage.com

This book is a translation from the original published under ISBN 978-620-5-50109-2.

Publisher:
Sciencia Scripts
is a trademark of
Dodo Books Indian Ocean Ltd. and OmniScriptum S.R.L publishing group

120 High Road, East Finchley, London, N2 9ED, United Kingdom
Str. Armeneasca 28/1, office 1, Chisinau MD-2012, Republic of Moldova, Europe
Printed at: see last page
ISBN: 978-620-5-60983-5

Table of Contents

Patent protection for integrated technical systems containing subsystems interconnected by artificial intelligence elements and artificial neural networks

Application of TRIZ and ARIZ analytical tools in drafting this type of patent application in the USA

The integration of software solutions and mobile applications into innovative technical systems at all levels, in today's process of transforming classic technologies into smart technologies and transforming the entire hierarchy of technical systems and their interconnections into smart technical systems, is the basis for the formation of just such most market-demanded innovative products.

As the practice of such innovative development shows, the use of methods and techniques of the inventive process adapted to today's element base and modern composite materials, from the generation of an idea to the complete formation of application materials for an invention, can be successfully conducted with the well-known Theory and Algorithm of Inventive Problem Solving.

The use of TRIZ and ARIZ is of course only one of many possible analytical tools when drafting a patent application.

Of particular importance in the innovation process are relationships with investors and potential strategic partners in the commercial application of the innovative product being created.

In order to convince an investor that investing in the proposed project will bring them financial success, it is necessary to explain the project in detail to its experts and disclose all the technological and innovative secrets.

This cannot be done without serious protection of the technical solutions underpinning the innovative product at all stages of project development.

Non-traditional types and forms of collective investment have now emerged and are being successfully developed, and this also forces project creators to constantly look for more and more secure forms of protection for their developments.

A provisional patent application is both an inexpensive and

secure option for up to 1 year of protection.

The application of TRIZ and ARIZ, as mentioned above, with the necessary and sufficient updating of basic techniques and methods, is only one of many possible tools in drafting a patent application and in many cases requires systematic modification and optimisation to take account of new innovative circumstances.

A company applicant or individual inventor, on a case-by-case basis, taking into account all the circumstances, reasons, objectives and conditions, taking into account the results of patent anticipation, in full compliance with the current US patent law (America Invents Act), can be, and more often than not, requires an integrative program, in which the original system and structure of the patent application is extremely precisely developed.

A patent licensing strategy, of which a consistent patent anticipation process is a part, can also be, and most often should be, developed for each individual applicant or inventor.

These instructions are also fully applicable to the preparation of the NON Provisional Patent Application (Utility Patent Application).

Apparatus and method for ...

The title of the proposed invention must contain not only a brief definition of the nature of the technical solution but also a very brief commercial characterisation of its market impact within the scope of the proposed technical solution. The title should not contain any advertising characteristics such as efficient, best, etc.

It is very important that the title of the invention takes account of the market situation at the time of filing the application. If the objectives of the authors and/or their companies and financial partners or applicants are to attract further investment in the development of the project for which the technical solution to be applied for is the basis, it is advisable to take into account the known technological and commercial interests of the potential partners in the title.

As evidence of non-obviousness of a claimed technical solution is very important to the US patent examiner, it is crucial to consider the criteria and nature of non-obviousness in the title of the invention.

Technical systems relating to so-called smart technology and equipment should normally include software that is not itself a technical solution and cannot be classified as a technical system.

But since software solutions are the basis of the control and monitoring systems of any modern smart technology, the name of the invention should also take this into account, i.e. the name of the invention should be in the form of "Apparatus, Programme, System and Method" or "Apparatus, Programme, System and Method".

In addition, when classifying a technical system as an intelligent system with elements of artificial intelligence and artificial neural networks, it is also necessary to have information stores with an appropriate amount of information and appropriate speed.

This factor should be reflected in the name, e.g. "Apparatus, Programme, System with Memory Stores" and the associated method of use

Reference to Related Applications

[0001] This application claims the priority benefit under 35 U.S.C. §119(e) of U.S. Provisional Application Nos. 61/, filed; and 61/, filed. The entire disclosures of each of these applications are incorporated herein by reference.

This section publishes information on all provisional patent applications that have been filed by the inventor or applicant prior to the making and filing of the present application.

Where no provisional patent application has been filed in the United States (Provisional Patent Applications), this section of the application provides information on all previous applications on the subject matter and the technological field that were filed by the authors of the invention or by the applicant (on behalf of the authors of this application or on behalf of other authors) before the present application was made and filed.

Technical Field

[0002] The invention relates to apparatus and methods for...

The nature of the field of technology to which the claimed technical solution pertains must be clearly substantiated and related to the distinctive features of the claimed technical solution by a cause-effect relationship described and formulated in a manner comprehensible to an average person skilled in the art.

It is particularly important to characterise the problems and contradictions which are characteristic and evident in the technological field selected for the invention and in the development of this thesis for the field of engineering and technology to which the claimed technical solution relates.

In view of the fact that, unlike the European Patent Agency, in the United States a patent can also be granted for a computer program if certain conditions are met, it is desirable to have an integration solution for the claimed technical solution in the type "Program, system, method and apparatus for its application".

Description of the Related Technology

[0003]

This section consists of two parts; In the first part, a more specific, local analytical assessment of the field of engineering and technology to which the claimed technical solution relates is given; A description of the most characteristic features of the technology, its known disadvantages and advantages is given.

[0004]

In this part of the section, it is advisable to begin a search for, and preliminary analysis of, controversies of a general nature which are common knowledge and characteristic of the technology to which the authors of the invention attribute it.

BACKGROUND OF THE INVENTION

[0005]

This section, which is one of the most important parts of the application, provides an in-depth analysis of the results of the patent search on all variants and sources of published patent information, at least the US Patent Office, preferably the European Patent Office, and even better a comparative analysis of known technical solutions and finally a structural element by element analysis.

After that, using TRIZ and ARIZ tools, it is necessary to search, formulate and compare all kinds of contradictions in known technical solutions and define and characterise a possible ideal final result.

Then analyse the possibility in the claimed technical solution to achieve the ideal end result in principle using techniques, technology, materials and means of control and management known at the time of filing the application.

Summary

[0006] According to some embodiments of the invention... this section defines the first local final aim of the invention and defines the manner in which it is to be achieved in accordance with the claims; it identifies the local aim and determines whether the local aim is sufficiently attained to constitute at least a partial or local ideal final result.

[0007] In some embodiments, programs, systems and methods of... this section defines the second local final aim of the invention and determines how to achieve it in accordance with the claims; it identifies the local aim and determines whether the local aim is sufficiently attained to constitute at least a partial or local ideal final result.

[0008] In some embodiments, further programs, systems and methods for... this section defines the third local final aim of the invention and defines the path to be taken to achieve it in accordance with the claims; it identifies the local aim and determines whether the local aim is sufficiently attained to constitute at least a partial or local ideal final result.

[0009] Also in some further embodiments, programs, systems and methods of... this section defines the first integral local end goal of the invention and determines the path to be taken to achieve it in accordance with the claims; identifies the integral local goal and determines whether the set goal is sufficient to constitute at least a partial or local idealized end result.

[0010] In some embodiments, an apparatus is provided with... This section defines the first local end purpose of the invention, which is causally related to the second local end purpose of the invention and defines a way of achieving said target integration within the scope of the claims; identifies the local integrated purpose and determines whether the local integrated purpose is sufficient to constitute at least a partial or local ideal end result.

[0011] This section defines the first local final aim of the invention in a causal relationship with the second and third local final aim of the invention and defines the way in which the said target integration can be achieved in terms of the correspondence between the claims; it identifies the local integrated aim and determines whether the achievement of that aim is sufficient to constitute at least a partial or local idealised final result.

[0012] In some embodiments, an apparatus for... This section defines the first local final aim of the invention in relation to the apparatus solutions, which is causally related to the second local final aim of the invention also related to the apparatus and defines the way to achieve said target integration within the scope of the claims; identifies the local integrated aim and determines whether the local integrated aim is sufficient to constitute at least a partial or local ideal

[0013] In some other embodiments, an apparatus is provided with a system for... in this section, one of the known systems required for its declared duty cycle is added to the apparatus and analysed for the presence or absence of a jump effect.

[0014] In some other embodiments, an apparatus is provided with a system for... In this section, the second known system that is required for its declared duty cycle is added to the apparatus and the presence or absence of a jump effect is analysed.

[0015] In some embodiments of the various apparatus, the... in this section, a combination of known systems integrated with the apparatus into an integrated system that is required for its declared duty cycle is added to the apparatus and the presence or absence of a jump effect is analysed.

[0016] In this section, the apparatus is added one by one to the combination of known systems required for its declared duty cycle and analysed for the presence or absence of a jump effect.

[0017] The system for... This section conducts a structural analysis of the system(s) of the technical solution and monitors whether the technical solution contains a sufficient combination of all the necessary systems to ensure that the ideal end result is achieved.

[0018] Structurally, the apparatus can be provided with a... In this section, a structural analysis of the components and assemblies of the apparatus is carried out.

[0019] In the apparatus, the system for... in this section show how design features influence the achievement of the ideal end result.

[0020] The apparatus can also comprise a system for... this section considers the possibility of incorporating various additional functionally independent and self-contained systems into the apparatus, which can enhance the effect of the proposed invention and thereby provide a more certainty of achieving the ideal end result.

[0021] In some embodiments, the method of operation of the device involves... this section analyses local and integral goals in dynamics or action, analyses the methods of achieving the goals and compares the parameters after the specified goal is achieved with the ideal end result.

[0022] An external... This section looks at additional possible aspects of improving and modifying the objectives and how they can be achieved, analysing and assessing the potential for achieving additional components of the ideal outcome.

[0023] The following can be provided to... In this section, various actions and design improvements that can add new content to the essence of the ideal end result are evaluated.

[0024] In some embodiments,... this section definitively defines the fully integrated local final aim of the invention and defines the path to be taken to achieve it in accordance with the claims, identifies the local, fully integrated aim and determines whether the aim is sufficiently achieved to constitute at least a partial or local ideal final result.

Brief Description of the Drawings

[0025] The invention will be better understood from the Detailed Description and from the appended drawings, which are meant to illustrate and not to limit the invention. The Figures are not necessarily drawn to scale, nor are the relative sizes of parts within the Figures necessary in proportion to one another.

[0026] Figure 1 is an example of a ..., - this figure presents a general schematic block diagram of a technical super-system, in which all subsystems and their links among themselves and outputs and communication lines with the supersystem are presented, focusing on functional features and differences of software elements integrated into the supersystem in combination with control and supervision systems of the supersystem containing elements of artificial intelligence and artificial neural networks.

[Figure 2 is an example of a ..., - this figure presents a general principle block diagram of a master and central technical system - subsystem, in which all functionally related subsystems and their links among themselves and outputs and communication lines with the supersystem are presented, focusing on functional features and differences of software elements integrated into the supersystem and, if possible, into subsystems in combination with the control and monitoring systems of the supersystem containing elements of artificial intelligence and artificial neural networks The technical details and software functions and algorithms are also reflected, which describe the non-obviousness of these solutions in combination with the general principles of the application of elements of artificial intelligence and artificial neural networks.

[0028] Figure 3 shows an example ..., - this figure presents a general principle block diagram of an auxiliary or local first technical subsystem, in which all functionally related subsystems and their links between them and outputs and lines of communication with the subsystem are presented, focusing on the functional features and differences of the software elements integrated into the subsystem and, where possible, into subsystems in combination with control and monitoring systems of the subsystem containing elements of artificial intelligence and artificial

[0029] Figure 4 shows an example of components of the ..., - this figure presents a general conceptual block diagram of an auxiliary or local second technical system - subsystem, in which all functionally related subsystems and their links with each other and outputs and links with the supersystem are presented, focusing on the functional features and differences of the software elements integrated into the subsystem and possibly into the subsystems, together with control and monitoring systems of the subsystem containing elements of artificial intelligence and claim

[0030] Figure 5 shows an example of a ..., - this figure presents a general principle block diagram of an auxiliary or local third technical subsystem, in which all functionally related subsystems and their links among themselves and outputs and communication lines with the supersystem are presented, focusing on functional features and differences of software elements integrated into the supersystem and, if possible, into subsystems in combination with control and monitoring systems of the supersystem containing elements of artificial intelligence and artificial neural networks. Technical details and software functions and algorithms are also reflected, characterising the non-obviousness of these solutions in combination with the general principles of application of elements of artificial intelligence and artificial neural networks.

[0031] Figure 6 is a, - illustrations to Figure 1 made as dependent graphical structures to the dependent claims relating to independent claim 1.

[0032] Figure 7 shows an example of a, - illustrations of Figure 2 made as dependent graphical structures to the dependent claims of independent claim 2.

[0033] Figure 8 is an example of a, - illustrations of Figure 3 made as dependent graphical structures to the dependent claims of independent claim 3.

[0034] Figure 9 is an example of a, - illustrations of Figure 4 made as dependent graphical structures to the dependent claims of independent claim 4.

[0035] Figure 10 is an example of a, - illustrations of Figure 5 made as dependent graphical structures to the dependent claims relating to independent claim 5.

[0036] Figure 11 is an example of a ..., - this figure presents a general principle block diagram or interaction algorithm of an auxiliary or local first technical system - subsystem, in which all functionally related subsystems and their links with each other and outputs and links with the supersystem are presented, focusing on the functional features and differences of the software elements integrated into the subsystem and possibly into the subsystem, together with control and monitoring systems of the subsystem containing elements artificially

[0037] Figure 12 is an example of an ..., - this figure presents a general principle block diagram or interaction algorithm of an auxiliary or local second-to-last technical system-subsystem, in which all functionally related subsystems and their links to each other and outputs and links to the supersystem are presented, focusing on the functional features and differences of the software elements integrated into the subsystem and, where possible, into the subsystem, together with the control and monitoring systems of the subsystem containing the element

[0038] Figure 13 is an example of a ..., - this figure presents a general principle block diagram or interaction algorithm of an auxiliary or local third-priority technical subsystem, in which all functionally related subsystems and their links to each other and outputs and links to the supersystem are presented, focusing on the functional features and differences of the software elements integrated into the subsystem and possibly into the subsystem, together with the control and monitoring systems of the subsystem containing the element

[0039] Figure 14 is an example of a ..., - this figure presents a general principle block diagram or interaction algorithm of an auxiliary or local technical subsystem, in which all functionally related subsystems and their links to each other and outputs and links to the supersystem are presented, focusing on the functional features and differences of the software elements integrated into the subsystem and possibly into the subsystem, in combination with the control and monitoring systems of the subsystem containing elements

[0040] Figure 15 is an example of a ..., - this figure presents a general principle block diagram or interaction algorithm of an auxiliary or local fifth-order technical subsystem, in which all functionally related subsystems and their links to each other and outputs and links to the supersystem are presented, focusing on the functional features and differences of the software elements integrated into the subsystem and possibly into the subsystem, together with the control and monitoring systems of the subsystem containing elements

[0041] Figure 16 is an example of a, - illustrations of figure 11 made as dependent graphical structures to the dependent claims of independent claim 11.

[0042] Figure 17 shows an example of, - illustrations to Figure 12 made as dependent graphical structures to the dependent claims of independent claim 12.

[0043] Figure 18 is an example of a, - illustrations of Figure 13 made as dependent graphical structures to the dependent claims of independent claim 13.

[0044] Figure 19 is an example of a, - illustrations of Figure 14 made as dependent graphical structures to the dependent claims of independent claim 14.

[0045] Figure 20 is an example of a, - illustrations of Figure 15 made as dependent graphical structures to the dependent claims of independent claim 15.

[0046] Figure 21 is an example of a ..., - this figure presents a general principle flowchart or algorithm for integrated integration into a central supersystem - an auxiliary or local first-in-class technical subsystem, in which all functionally related subsystems and their connections to each other and outputs and communication lines to the subsystem are represented, focusing on the functional features and differences of the software elements integrated into the subsystem and, where possible, into subsystems in combination with control systems and counter Technical details and software functions and algorithms describing the non-obviousness of these solutions in combination with the general principles of application of artificial intelligence elements and artificial neural networks are also reflected.

[0047] Figure 22 is an example of a ..., - this figure presents a general principle flowchart or algorithm for integrated integration into a central supersystem - an auxiliary or local second-best technical subsystem, in which all functionally related subsystems and their links to each other and outputs and links to the supersystem are presented, focusing on the functional features and differences of the software elements integrated into the supersystem and, where possible, into the subsystems in conjunction with control and monitoring systems Technical details and software functions and algorithms are also reflected, characterizing the non-obviousness of these solutions in combination with the general principles of application of artificial intelligence elements and artificial neural networks.

[0048] Figure 23 is an example of a ..., - this figure presents a general principle flowchart or algorithm for integrated integration into a central supersystem - an auxiliary or local third-priority technical subsystem, in which all functionally related subsystems and their links among themselves and outputs and communication lines to the supersystem are represented, focusing on the functional features and differences of the software elements integrated into the supersystem and, where possible, into the subsystems in combination with control systems and counter Technical details and software functions and algorithms describing the non-obviousness of these solutions in combination with the general principles of application of artificial intelligence elements and artificial neural networks are also reflected.

[0049] Figure 24 is an example of a ..., - this figure presents a general principle block diagram or algorithm for the complex integration into a central subsystem of an auxiliary or local quadruple technical subsystem, in which all functionally related subsystems and their links to each other and outputs and links to the subsystem are presented, focusing on the functional features and differences of the software elements integrated into the subsystem and, where possible, into subsystems in combination with control and control systems

detailed description

[0050] According to some embodiments, this section shows the elements of substantial novelty contained in the envisaged invention at the level of the basic technical solution at the top of the hierarchy of new or updated technical solutions and integrative combinations thereof.

[0051] In some embodiments, a, - this section shows the principles of integrating software solutions in a basic technical solution.

[0052] In order to develop, - this section shows the principle elements of the development of all types of technical solutions at the level of the basic technical solution without integration with software and control solutions.

[0053] To develop a, - this section shows methods and processes and algorithms designed for deep integration of software solutions of all kinds in the complex and hierarchical structure of the main technical solution and its integrative combinations with technical subsystems.

[0054] It will be appreciated that some, - this section shows the linkage and consequence of software solutions to the fundamental characterisation and relationships in all interconnected and mutually integrated technical systems, taking into account the essential elements of artificial intelligence and associated artificial neural networks.

[0055] In some embodiments, the, - in this section, the characteristics and new features that have emerged when integrating all levels of software solutions into the core technical solution.

[0056] It will be appreciated that the, - this section shows the elements of substantial novelty that have emerged when integrating software solutions into the classic type of technical solutions at all levels of the hierarchy.

[0057] In some embodiments, the, - this section gives the detailed characteristics and working relationships of the elements of artificial intelligence to the basic scheme of operation of technical systems at all levels of the hierarchy, linking the principles of substantial novelty to the algorithm or operating principles of the entire complex invented technical system.

[0058] In some embodiments, an, - this section gives the detailed characteristics and working relationships of the elements of artificial intelligence and artificial neural networks to the basic scheme of operation of technical systems at all hierarchical levels, linking the principles of substantial novelty to the algorithm or operating principles of the entire complex invented technical system.

[0059] Reference will now be made to the figures, in which like numerals refer to like parts throughout.

[0060] In Figure 1... This section gives a detailed description of Figure 1 with an explanation of all symbols and item numbers or numerical symbols; all numerical symbols for this figure should be of the form, - 101, 102... The following reference numerals identify the following features:

[0061] Figure 2 shows an example of a, - This section gives a detailed description of Figure 2, together with an explanation of all symbols and item numbers or numerals; all numerals for this figure shall be in the form, - 201, 202 ... The following reference numerals identify the following features:

[0062] Figure 3 shows a, - This section gives a detailed description of Figure 3 together with an explanation of all symbols and item numbers or numerals; all numerals for this figure shall be in the form, - 301, 302. The following reference numerals identify the following features:

[0063] Figure 4 shows a, - This section gives a detailed description of Figure 4, together with an explanation of all symbols and item numbers or numeric symbols; all numeric symbols for this figure should be of the form, - 401, 402. The following reference numerals identify the following features:

[0064] Figure 5 shows a, - This section gives a detailed description of Figure 5, together with an explanation of all symbols and item numbers or numeric symbols; all numeric symbols for this figure should be of the form, - 501, 502. The following reference numerals identify the following features:

[0065] Figure 6 shows a, - This section gives a detailed description of Figure 6, together with an explanation of all symbols and item numbers or numerals; all numerals for this figure should be of the form, - 601, 602. The following reference numerals identify the following features:

[0066] Figure 7 shows, - This section gives a detailed description of Figure 7 together with an explanation of all symbols and item numbers or numeric symbols; all numeric symbols for this figure should be of the form, - 701, 702. The following reference numerals identify the following features:

[0067] Figure 8 shows an, - This section gives a detailed description of Figure 8 with an explanation of all symbols and item numbers or numeric symbols; all numeric symbols for this figure should be of the form, - 801, 802. The following reference numerals identify the following features:

[0068] Figure 9 shows a, - This section gives a detailed description of Figure 9, together with an explanation of all symbols and item numbers or numeric symbols; all numeric symbols for this figure shall be of the form, - 901, 902 ... The following reference numerals identify the following features:

[0069] Figure 10 shows a, - This section gives a detailed description of Figure 10, giving an explanation of all symbols and item numbers or numerals; all numerals for this figure should be of the form, - 1001, 1002. The following reference numerals identify the following features:

[0070] Figure 11 shows, - This section gives a detailed description of Figure 11 together with an explanation of all symbols and item numbers or numeric symbols; all numeric symbols for this figure should be of the form, - 1101, 1102. The following reference numerals identify the following features:

[0071] Figure12 shows, - This section gives a detailed description of Figure 12 together with an explanation of all symbols and item numbers or numeric symbols; all numeric symbols for this figure should be of the form, -1201, 1202. The following reference numerals identify the following features:

[0072] Figure 13 shows a, - This section gives a detailed description of Figure 13, together with an explanation of all symbols and item numbers or numerals; all numerals for this figure shall be in the form, - 1301, 1302. The following reference numerals identify the following features:

[0073] Figure 14 shows, - This section gives a detailed description of Figure 14 together with an explanation of all symbols and item numbers or numeric symbols; all numeric symbols for this figure should be of the form, - 1401, 1402. The following reference numerals identify the following features:

[0074] Figure 15 shows a, - This section gives a detailed description of Figure 15, together with an explanation of all symbols and item numbers or numeric symbols; all numeric symbols for this figure shall be of the form, - 1501, 1502 ... The following reference numerals identify the following features:

[0075] Figure 16 shows a, - This section gives a detailed description of Figure 16, giving an explanation of all symbols and item numbers or numerals; all numerals for this figure should be of the form, - 1601, 1602. The following reference numerals identify the following features:

[0076] Figure 17 shows, - This section gives a detailed description of Figure 17 together with an explanation of all symbols and item numbers or numerals; all numerals for this figure shall be in the form, - 1701, 1702. The following reference numerals identify the following features:

[0077] Figure 18 shows a, - This section gives a detailed description of Figure 18 with an explanation of all symbols and item numbers or numeric symbols; all numeric symbols for this figure should be of the form, - 1801, 1802. The following reference numerals identify the following features:

[0078] Figure 19 shows a, - This section gives a detailed description of Figure 19, giving the meaning of all symbols and item numbers or numerals; all numerals for this figure shall be in the form, - 1901, 1902. The following reference numerals identify the following features:

[0079] Figure 20 shows, - This section gives a detailed description of Figure 20 together with an explanation of all symbols and item numbers or numerals; all numerals for this figure shall be in the form, - 2001, 2002. The following reference numerals identify the following features:

[0080] Figure 21 shows a, - This section gives a detailed description of Figure 21, together with an explanation of all symbols and item numbers or numerals; all numerals for this figure shall be in the form, - 2101, 2102. The following reference numerals identify the following features:

[0081] Figure 22 shows an, - This section gives a detailed description of Figure 22, together with an explanation of all symbols and item numbers or numerals; all numerals for this figure shall be in the form, - 2201, 2202 ... The following reference numerals identify the following features:

[0082] Figure 23 shows a, - This section gives a detailed description of Figure 23, together with an explanation of all symbols and item numbers or numeric symbols; all numeric symbols for this figure shall be of the form, - 2301, 2302 ... The following reference numerals identify the following features:

[0083] Figure 24 shows a, - This section gives a detailed description of Figure 24, together with an explanation of all symbols and item numbers or numeric symbols; all numeric symbols for this figure shall be of the form, - 2401, 2402 ... The following reference numerals identify the following features:

TEST RESULTS

[0084] ... This section shows the test results in the form of indicators, tables, graphs and charts.

[0085] Test results showed that in the... in this section, the test results given are identified and compared with similar test results of known devices or apparatuses.

[0086] ... in this section prove that the test results obtained are in line with the expected and declared indicators of the ideal end result.

ANALYSIS OF THE TEST RESULTS

[0087] The qualification test result analysis showed that the embodiments of the invention achieve the objective of... this section provides a structural analysis and characterisation of the test results of the prototypes of the invented technical solution.

EXAMPLES OF APPLICATIONS

[0088] In some embodiments, -... (examples are given of various embodiments or various embodiments of the invention, if any; if a variety of embodiments and a variety of applications are not included in the main distinctive features of the claims, it is essential that these are described in as clear and comprehensible detail as possible; Further, when identifying and verifying the actual use of the invention, these designs and examples of use help to provide greater evidence of the use of the invention

[0089] In some embodiments, -... (examples are given of various embodiments or various embodiments of the invention, if any; if the variety of embodiments and the variety of applications are not covered by the main features of the claims, it is essential that these are described in as clear and comprehensible detail as possible; Further, when identifying and verifying whether the invention is actually used, these designs and examples of applications help to provide greater evidence of the use of the invention.

[0090] In some embodiments,... it is stated which original features and combinations of features in each embodiment considered are responsible for achieving the ideal end result expected of that particular embodiment.

[0091] In some embodiments,... it is stated which original features and combinations of features in each embodiment considered are responsible for achieving the ideal end result expected of that particular embodiment.

[0092] In some embodiments,... it is stated which original

features and combinations of features in each embodiment considered are responsible for achieving the ideal end result expected of that particular embodiment.

[0093] In some embodiments,... it is stated which original features and combinations of features in each embodiment considered are responsible for achieving the ideal end result expected of that particular embodiment.

[0094] In some embodiments,... it is stated which original features and combinations of features in each embodiment considered are responsible for achieving the ideal end result expected of that particular embodiment.

[0095] In some embodiments,... it is stated which original features and combinations of features in each embodiment considered are responsible for achieving the ideal end result expected of that particular embodiment.

[0096] In some embodiments,... it is stated which original features and combinations of features in each of the embodiments considered are responsible for achieving the ideal end result expected of that particular embodiment.

[0097] In some embodiments,... it is stated which original features and combinations of features in each of the embodiments considered ensure that the ideal end result expected of that particular embodiment is achieved.

[0098] In some embodiments,... it is stated which original features and combinations of features in each embodiment considered are responsible for achieving the ideal end result expected of that particular embodiment.

[0099] An operating principle of the devices is... The options for operating the apparatus or device are analysed and compared, options for control processes, monitoring and the application of principles for the formation of feedback elements between actuators and systems or control processors are analysed.

[0100] ... The possibilities and features of the various embodiments and examples of applications of the invention listed in the section are compared.

[0101] In some embodiments,... It is stated which original features and combinations of features in each of the embodiments considered ensure that the ideal end result expected of that particular embodiment is achieved.

[0102] In some embodiments,... indicates which original features and combinations of features in each embodiment considered ensure that the ideal end result expected of that particular embodiment is achieved.

[0103] It will be appreciated that the... indicates which original features and combinations of features in each of the variants considered ensure that the ideal end result expected of that particular performance is achieved.

[0104] Various embodiments of the invention allow one or more of the following advantages:

... Any positive results arising from the implementation of the invention are indicated and analysed.

[0105] It will be appreciated by those skilled in the art that various omissions, additions and modifications may be made to the methods and structures described above without departing from the scope of the invention. All such modifications and changes are intended to fall within the scope of the invention, as defined by the appended claims.

WHAT IS CLAIMED IS:

1. The first independent claim is the most important for the application;

The first claim should be as concise as possible but in three parts.

The first part gives the commercial name of the technical solution, which should cover as wide a field of commercial use as possible while giving an indication of the scope of the claim of the technical solution claimed, excluding advertising language and paradoxical statements.

The second part of the first paragraph describes the distinctive features, their combinations and relationships that are known and used as the basis for realising the distinctive features.

The third part of the first paragraph describes the distinctive features, their combinations and the static and dynamic relationship of the distinctive features to the basic known features, which together ensure that the ideal end result is achieved.

Subsequent dependent claims should set out the distinctive feature in as much detail as possible in general terms. All subsequent claims would disclose all possible uses of the distinctive feature to generate an effect and achieve the ideal end result. If the ideal end result can be achieved in some variations and subject to a variety of possible combinations of the distinctive features with the basic features, all the possible variations are indicated in the following claims.

Each subsequent claim also consists of three parts, each of which complies with the requirements of the first claim

- Second independent claim
- Dependent claim of the first and third claims
- Dependent claim of the first and third claims
- Dependent on the first and third claims
- Dependent on the first and third claims
- Third independent claim

If the subject matter of the invention is an apparatus or apparatus and is named, for example, "Apparatus...", it is desirable to have at least three independent claims, the first claim being "Apparatus for..., method of use and associated method of manufacture".

The third point is the method of application of the apparatus....

The eighth item is "Apparatus for... and the associated method of making it".

The above is only an example, each technical solution has its own specific characteristics and the composition and structure of the claims may vary accordingly.

Abstract of the Disclosure

The abstract is in principle unremarkable and does not require any guidelines on formatting.

Composite technical solutions as the basic basis for the formation of innovative technical systems and their constituent software products as effective equivalents for complex integrative inventions

Composite technical solutions, especially those containing fragments and components that constitute integrative inventions, have recently been increasingly referred to as smart technology, smart materials, smart manufacturing, etc.

This peculiar understanding, started with a definition - smart materials.

There is, of course, no formal definition of the catchy term 'Smart Materials'. It usually refers to materials that are able to change their properties under the influence of their environment. For example, there are so-called shape memory materials.

For example, a wire made from a nickel-titanium alloy (also called a shape memory alloy), when bent, will return to its original shape after being heated.

These properties were first used in medical technology as far back as the last century, when performing operations to close the Batalov's ductus septum between the ventricles of the heart.

An umbrella of nickel-titanium alloy micro-plates was created on which an elastic sheath of silicone rubber was stretched. The umbrella, when folded, was fed through the catheter to the point where there was an open duct, and when the body temperature was reached, shape memory returned the nickel-titanium alloy plates to their original position. The umbrella closed the duct and the heart defect was closed without surgery.

It should be noted straight away that, despite the freshness of the term itself, the effects associated with smart materials were discovered quite a long time ago.

For example, the shape memory effect was studied as early as the 1930s, and the properties of nickel and titanium alloys were studied by Soviet metallurgists Kurdyumov and Khandorson in 1948 (although the alloy got its name because the Americans rediscovered it in the 1960s, but that's another story).

This text deals with a special class of smart materials - self-healing materials. The term refers to technical systems that are capable of resisting structural failure due to mechanical impact.

The main requirement for such materials is that the damage must be "Healed" without human intervention. There are a great many mechanisms for such healing.

One way of protecting metal (e.g. aluminium) parts is by coating them with chrome and treating them with special solutions, often containing chromic acid.

This treatment forms a thin layer that protects the metal against corrosion. This layer also serves as an excellent primer for subsequent painting or spraying. Chrome plating, however, uses hexavalent chrome. This material is considered hazardous to health because, unlike, for example, trivalent chromium, it penetrates living cells relatively easily.

However, some industries, such as the aerospace industry, still use hazardous chrome plating on aluminium alloys.

Among other things, this is due to the fact that, when in use, such a coating is able to heal small scratches and damage on its own within a few weeks. Roughly speaking, the chrome itself migrates to the scratched area, filling and covering it. Of course, the scratch will not heal during this migration (the product's marketable appearance will still be ruined), but the aluminium alloy substrate will be protected (worse, of course, but protected).

Thus, one of the tasks facing chemists is to create a coating that could replace the dangerous chrome plating. At the end of the year, scientists from the University of Nevada presented their prototype of such a coating. They came up with a molybdenum-based coating. The scientists also proposed a way to apply this coating to the surface of AA2024-T6, an aluminium alloy used in the space industry.

During the study, the scientists intentionally damaged the sample. Then, using multiple spectroscopic techniques (to be sure), they verified that molybdenum was present in the damaged area, meaning that the coating is capable of self-repair. The scientists themselves say that their work is not yet complete - they are working on improving the formulation of the coating. Remarkably, the researchers tried about 200 different formulations before getting acceptable results.

But smart materials and technologies, today, are primarily technical systems with complete recycling of consumables and no waste.

One of the most important processes in modern chemistry and electrochemistry is the process of continuous regeneration in real time, e.g. of pickling solutions in the production of electronic and microelectronic boards.

When designing these kinds of composite complex processes, it is always the process of correcting the acidity or alkalinity level of the solution to be regenerated that is most critical, without the use of chemicals.

In the last decade, methods and apparatus have been invented for such a purpose.

Until today, this integrative invention has never been improved or anyone has been able to create something better.

Recently, composite composite technical systems have also appeared, the evolution of which is quite dynamic, and at each stage of development all used technical systems represent - complex integrative inventions, having in their composition a number of successive elements of structural and technological regeneration, tied to the functional algorithm of the supersystem.

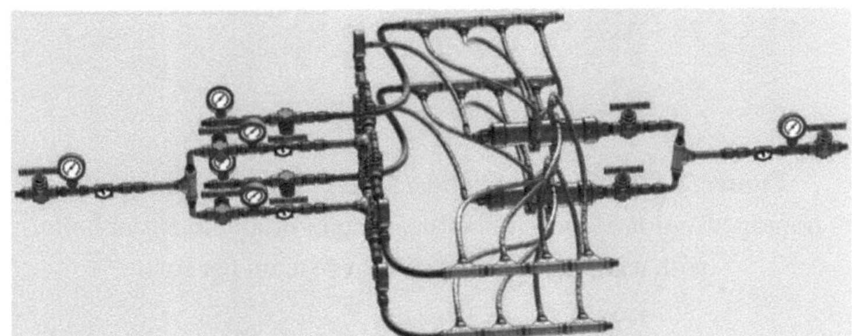

Figure 1: Duplicated technical system for a liquid fuel line homogenisation station in the fuel system of an industrial boiler, with a capacity of 10 tonnes of steam per hour

The technical system is based on standard component parts,

without composites and without reference to the production capabilities of digitally controlled machines.

As can be seen in the figure, the technical system also includes monitoring gauges to monitor pressure and flow in real time, but with control from valves operating in manual mode.

The homogenising devices are connected and work in parallel, complementing each other.

Elements and fittings of standard pipework have been applied for their main purpose, which required somewhat larger dimensions, while maintaining the required performance.

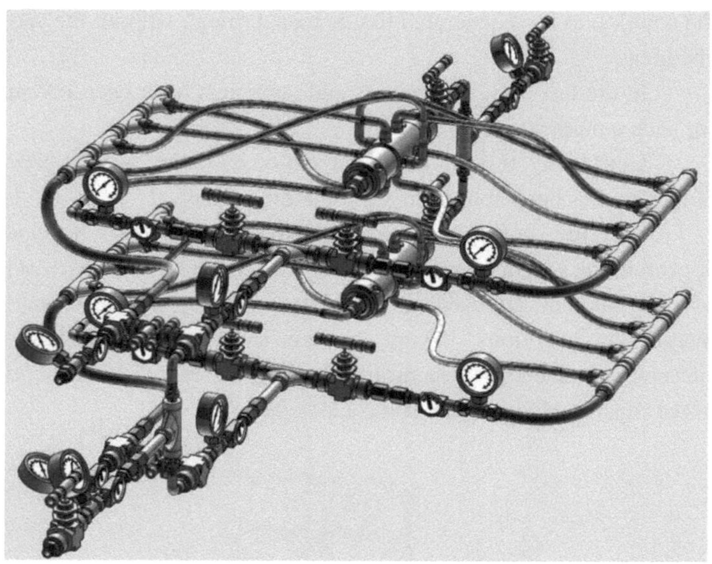

Figure 2: Duplicated technical system for a liquid fuel line homogenisation station in the fuel system of an industrial boiler, with a capacity of 10 tonnes of steam per hour

The technical system is based on standard component parts, without composites and without reference to the production capabilities of digitally controlled machines.

As can be seen in the figure, the technical system also includes monitoring gauges to monitor pressure and flow in real time, but with control from valves operating in manual mode.

The devices for homogenisation are connected and work in parallel, mutually complementing each other, but the larger scale of the pattern allows the characteristic colour of the copper and copper alloy piping elements to be seen, which in addition to the high cost results in the need for regeneration, because the stability, strength and durability of copper parts under high pressures and in the presence of chemically active substances have much lower values.

The need for a fundamental transformation of traditional technical systems into composite technical solutions with innovative structure and innovative functionalities and characteristics forming a vertical and horizontal hierarchy of integrative invention, leads to the need to redefine the techniques and methods of consistent transition from the basic composite foundation of an existing technical solution or complex technical system to a new technical solution and to the basic foundation and

Techniques and methods of transition from the established composite foundation of a new technical solution to the basic foundation of an integrative invention

As the first stage of technical system modification we should name - the stage of complex and consecutive replacement of system components by more advanced and adapted to work in the conditions of the composite technical and technological cycle.

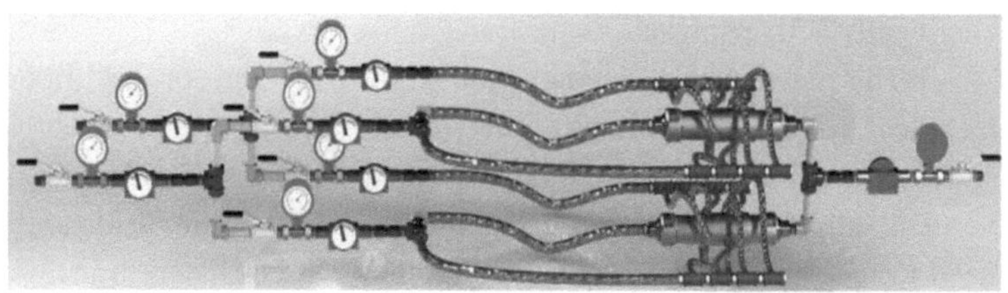

Figure 3: Upgraded redundant technical system for liquid fuel line homogenisation station in the fuel system of an industrial boiler, with a capacity of 10 tonnes of steam per hour

The technical system is based on innovative and modified standard components, with maximum use of composite materials and linked to the production capabilities of digitally controlled machines.

As can be seen in the figure, the technical system also includes monitoring gauges for real-time pressure and flow monitoring, but controlled by valves operating in semi-automatic mode.

The homogenising units are connected and operate in parallel, complementing each other, but have the option of operating separately in an independent operating mode if required.

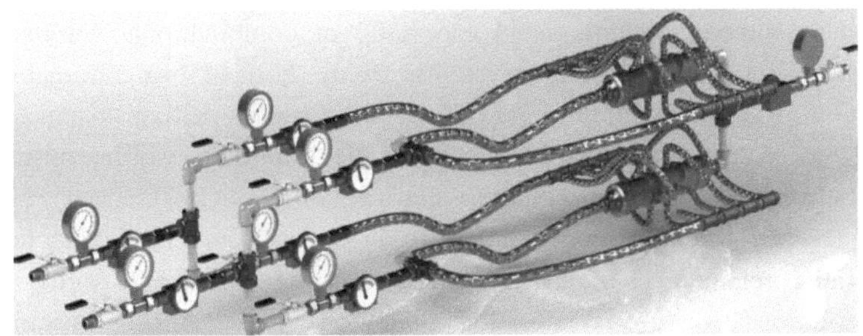

Figure 4: System-modernised redundant technical system for a liquid fuel line homogenisation station in the fuel system of an industrial boiler, with a capacity of 10 tonnes of steam per hour

The subsequent modification of the technical subsystem and the technical supersystem are based on standard, but modified component parts, using composite materials as much as possible and with structural and technological reference and adaptation to the production and technological possibilities of the digitally controlled equipment.

As can be seen from the figure, the technical system also includes automated control gauges to monitor pressure and flow in real time, but controlled by valves operating in semi-automatic mode and having the capability to be part of a closed-loop automatic control and monitoring system with programmable controllers.

The homogenisation devices are connected and work in parallel, complementing each other, but their technical capabilities and characteristics allow optimisation within the existing technical solution, taking into account the demands of the supersystem.

The need to move to combinatorial technical systems, especially those at the supersystem level, requires optimisation and fundamental directions for the development of TRIZ and ARIZ.

The following key areas of focus can be identified:

1. A traditional phenomenon in the evolution of TRIZ and ARIZ is a general increase in the degree of algorithmisation through fuller and deeper use of objective laws of technical system

development, including and especially of combinational complex technical systems (supersystems) with built-in programmable controllers and with more powerful control and modelling systems.

2. Substantial strengthening of the 'bridge' (causal inference) between physical contradictions and ways of resolving them.

3. Strengthening the pool of information, strengthening the links between TRIZ, ARIZ and current standards, and software solutions for pre-computation operations and real-time system modelling and simulation.

4. Transformation of the second half of ARIZ (development and use of the idea found) into a self-contained and independent algorithm using the latest software technology and advanced computer and processor technology capabilities.

5. Development of a new initial part (or a separate algorithm) to identify new tasks including innovative tasks with adaptation to the full simulation cycle.

6. Strengthening the general education and general training function on the basis of the latest techniques in system programming, modelling and simulation.

TRIZ and ARIZ should more vigorously and selectively develop the skills of concentrated, goal-oriented and logically strong thinking based on the advanced capabilities of modern computer technology.

7. Gradual integrative multi-level increase in compositional universality, compositional interchangeability, mutual selective applicability and modularity in the compositional format.

Alignment of functions and reallocation of functional load between the local modules of the innovation system.

It makes sense to elaborate on this perspective.

A sequential step-by-step algorithm is proposed for the implementation of this task, of course, in a multifunctional version with maximum unification and full local interchangeability of all links and elements of the innovative technical solution.

PART 1

1. FIRST STEP

Identify and formulate the final version of the idea generated and the ultimate goal of solving the problem of implementing that idea

What is the technical objective of the problem ("What characteristic of the object should be changed?") Are the parameters of the task related to those of innovative tasks that have already been successfully implemented?)

Which characteristics of an object cannot be knowingly changed when solving a problem?

What is the economic aim of solving the problem and what is the commercial aim ("What costs will be reduced if the problem is solved?")

What are the (approximate) acceptable costs?

What is the main technical and economic indicator that needs to be improved?

1- 2. SECOND STEP

Check the workaround. Suppose the problem is fundamentally unsolvable; what other - more general, known and already solved - problem should then be solved to get the required end result?

(Use a systems analysis - a framework for such an analysis could be based on the principles of the patenting and licensing strategy published earlier in the same resource).

1- 3. THE THIRD STEP

Determine whether it is more appropriate to solve the original problem or the workaround.

Compare the original task with the trends in the technology sector.

Compare the initial task with the trends in the leading technology sector.

Compare the bypass problem with the development trends in

the technology sector.

Compare the bypass problem with the development trends of the leading technology sector.

Match the original task with the bypass task. Make a selection.

1-   4. FOURTH STEP

Determine the quantitative indicators required.

1-   5. FIFTH STEP

Incorporate a "Time Amendment" into the required quantitative indicators.

1- 6. SIXTH STEP

Specify the requirements necessitated by the specific circumstances in which the invention is to be implemented.

Take into account the special features of the implementation. In particular, the degree of complexity allowed for the solution.

Take into account the intended scope of application.

PART 2. CLARIFYING TASK CONDITIONS

2- 1. FIRST STEP

Clarify the task using the patent literature. Search the US Patent Office for granted patents, patent applications and posted patent applications. Search the holdings of the European Patent Office under subheadings for names of authors of inventions, applicant companies for inventions and for composite indexes and keywords

How (according to patent data) are problems close to this one solved?

How are problems similar to this one solved in the leading technology sector?

How are the inverse of this problem solved?

2-   2. SECOND STEP

Apply the RVS operator.

a . We mentally change the size of the object from a given

value to 0 (P->0). How is the problem now solved?

b . We mentally change the size of the object from a given value to yes (P-> yes). How is the problem now solved?

at . We mentally change the process time (or the speed of the object) from a given value to 0 (B->0). How is the problem now solved?

r . We mentally change the time of the process (or the speed of the object) from a given value to yes (B-> yes). How is the problem now solved?

e . We mentally change the cost (allowable cost) of an object or process from a given value to 0 (C->0). How is the problem now solved?

e . We mentally change the cost (eligible costs) of an object or process from a given value to yes (C-> yes). How is the problem now solved?

2-   3. THE THIRD STEP

State the conditions of the task (without using special terms or specifying what you need to invent, find, create) in two phrases in the following form:

a . A system is given which includes elements (list).

b . Element (specify)

at .   subject to (specify)

d . has an undesirable effect (specify).

Examples:

"An electrical circuit is given, consisting of a generator, a conductor, a current consumer and a switch. The circuit breaker produces an electric arc when it opens and closes the circuit" (Problem 1).

"There is a pipeline with a gate valve; water with particles of iron ore is moving along the pipeline. The ore particles are rubbing the gate valve" (problem No. 21).

"A cove is given, consisting of banks, bottom, water and a layer of silt lying on the bottom. Elimination of silt using known methods causes pollution of the environment, requires a lot of time and money" (task P).

"Given a heavy concrete pipe standing vertically and a slope on which it has to be lowered. If the pipe is lowered without complex crane equipment, the pipe could collapse as it falls onto the slope" (Problem 37).

2-4. FOURTH STEP
Transfer the elements from 2-H to the following table:

	If you consider the system as a device	If you see the system as a way to
a. Elements that can be changed, reworked, reconfigured (in the context of this task).		
6. Elements that are difficult to modify (in the context of this task).		

Examples: In Problem No. 1, the generator, the conductor and the circuit breaker refer to 'a';

Current consumer - to 'b' (consumers are many off-the-shelf and different machines, appliances; they are difficult to rebuild).

In problem no. 21, the pipeline and gate valve are 'a', the water and ore particles are 'b'.

In Problem P, all the elements belong to 'b'.

In problem 37, the slope and pipe are "b" (the pipe cannot be changed at all, the slope can only be changed to a minor extent).

Note: one vertical row is sufficient, i.e. it is sufficient to consider the system as a device **or** as a method.

2- 5. FIFTH STEP

Select from 2-4a the element that is most amenable to change, remodelling, readjustment.

Note:

a . If all elements in 2-4a are equal in terms of the degree of change allowed, start with **the stationary** element (usually easier to change than the movable element).

b . If there is an element in 2-4a that is directly related to an undesirable effect (usually this element is indicated in 2-3 b), select it last.

at .   If the system only has elements 2-4b, take the external environment as an element.

Examples:

In Problem No.1, three elements belong to 2-4a - generator, conductor, circuit breaker. The circuit breaker is associated with an undesirable phenomenon (arcing). The generator or the conductor must be chosen (the generator is generally stationary, so the conductor cannot be preferred).

In problem no. 27, a pipeline must be chosen, as the gate valve is associated with an undesirable phenomenon (abrasion).

In problems P and 37, all elements belong to 2-4b, so the external environment should be chosen.

PART 3. ANALYTICAL STAGE

3- 1 FIRST STEP

Draw up the wording of the ICD in the following form:

a . Object (take the element selected in 2-5);

b . What does.

at .   As he does himself.

d . When does.

e . Under what mandatory conditions (restrictions, requirements, etc.)

Examples:

In task 1 - "The generator... interrupts and restores

current supply... itself... when the consumer needs it... without

arcing".

In problem no. 21 - "A pipeline ... changes its cross-section... by itself... when it is necessary to regulate the flow... without fraying."

In task P - "The external environment... eliminates il... itself... in a short time... without polluting the sea, coasts and atmosphere".

In Problem No. 37 - "The external environment... lowers the pipe... itself... during laying... without damage".

3-   2. SECOND STEP

Make two drawings: "was" (before IRR) and "became" (IRR).

Notes:

but . The drawings can be conventional - as long as they reflect the "was" and "became".

b . The "became" picture should match the verbal wording of the IRR.

Verification:

The drawings must contain all the elements listed in 2-3 a.

If an external environment is selected in step 2-5, it should be indicated in the "became" figure.

3-   3. THE THIRD STEP

In the "became" diagram, find the element indicated in 3-1a and highlight the part of it that cannot perform the required action under the required conditions. Mark this part (by hatching, using a different colour, outlining, etc.).

Examples:

In problem no. 21, this will be the **inside** surface of the pipeline.

In Problem No. 37, the section of the outside **between the pipe and the slope**.

In Problem P, the sub-bottom area of the bay (other areas of the external environment - offshore, onshore, atmospheric - are excluded by conditions 3-1d).

3-   4. FOURTH STEP

Why can't this part carry out the required action?
Examples:
In Problem No. 21 - "The inner surface of the pipe does not know how to compress and decompress".

In Problem No. 37 - "The external medium between the pipe and the slope does not know how to lower the pipe smoothly".

In Problem P - "The sub-bottom (deep, underground) area has no voids connected to the bottom of the bay.

3-5. FIFTH STEP

Under what conditions will this part be able to carry out the desired action (what properties should it have)?

Note: Don't think yet - whether the desirable property is practicable. Name this property without worrying about how it will be achieved.

Examples:
In Problem No. 21 - "Something builds up on the inner surface of the pipe and then - if necessary - disappears".

In Problem No. 37 - "The external medium between the pipe and the slope has a density sufficient for the smooth lowering of the pipe".

In Problem P - "Cavities communicating with the bottom of the bay have been created in the sub-bottom area".

3-6. SIXTH STEP

What should be done to make the selected part of the object acquire the properties marked in 3-5?

Figure:
Supporting questions:
a . Show in the figure with arrows the forces that must be applied to the selected part of the object in order to provide the desired properties.

b . In what ways can these forces be created?

(Cross out ways that violate conditions 3-1d).

Examples:

In Problem No. 21 - "Build up iron ore particles or water (ice) on the inside of the pipe". There are no other substances inside the pipe, this determines the choice.

In problem 37 - "Place a dense medium between the pipe and the slope instead of air, which would be easy to remove gradually - ice, sand, etc.".

In task P - "Find natural or create artificial cavities under the bottom of the bay, connect them with boreholes to the bottom".

3-7. SECOND STEP

Formulate a way that can be practically implemented. If there is more than one, mark them with a number (the most promising one with number 1, etc.). Write down the chosen methods.

3- 8. STEP 8.

Give a diagram of the apparatus for carrying out the first method:

Figure:

Supporting questions:

a . What is the aggregate state of the working part of the device?

b . How does the device change during one work cycle?

at .   How does the device change after many cycles?

(After solving the problem, go back to step 3-7 and consider the other methods listed there.)

PART 4. PRE-ASSESSMENT OF THE IDEA FOUND

4- 1. FIRST STEP

What is made worse by using the proposed device? Write down - what becomes more complicated, more expensive, etc.

4- 2. SECOND STEP

Can this deterioration be prevented by modifying the proposed device? Draw a diagram of the modified device.

4- 3. THE THIRD STEP

What is the deterioration now (what is more complicated, more expensive, etc.)?

4- 4. FOURTH STEP

Compare winning and losing:

a. What's bigger?

6. Why?

If the gain is greater than the loss (at least in perspective), move on to the synthetic part of ARIZ.

If the loss is greater than the gain, return to step 3-1. Record on the same sheet the course of the reanalysis and its result.

4-5. FIFTH STEP

If the gain is now greater, move to the synthetic ARIZ stage. If the reanalysis does not give any new results, go back to step 2-4, check the table. Take another element of the system in 2-5 and re-run the analysis. Record the course of analysis on the same sheet.

If there is no satisfactory solution after 4-5, move on to the next part of the ARIZ.

Influence of restrictions on the number of claims on the ability to reliably protect composite technical solutions

Proposed structure of an independent claim based on a composite technical solution

The first independent claim is the most important for the application;

The first claim should be as concise as possible but in three parts:

The first part gives the commercial name of the technical solution, which should cover as wide a field of commercial use as possible, while giving an indication of the scope of the claim of the technical solution claimed, excluding advertising language and paradoxical statements...

The second part of the first paragraph describes the distinctive features, their combinations and relationships that are known and used as the basis for realising the distinctive features;

Part three of the first paragraph describes the distinctive features, their combinations and the static and dynamic relationships of the distinctive features to basic known features which, taken together, provide the ideal end result

Proposed structure of a dependent claim based on a composite technical solution

Subsequent dependent claims should set out the distinctive features in as much detail as possible in a generic form; All subsequent claims should disclose all possible embodiments of the distinctive feature in order to achieve the ideal end result; If the ideal end result can be achieved in some variations and subject to different combinations of the distinctive features with the basic features, all possible variations are stated in the subsequent claims;

Each subsequent claim also consists of three parts, each of which complies with the requirements of the first claim

1. Second independent claim
2. Dependent claim   of the first and third claims
3. Dependent claim   of the first and third claims
4. Dependent on   the first and third claims
5. Dependent on   the first and third claims
6. Third independent claim

If the subject matter of the invention is an apparatus or an apparatus and is named, - for example, - apparatus..., it is desirable to have at least three independent claims, so that the first claim is, - apparatus for..., method of use and associated process for making...

The third point, the method of application of the apparatus ...;

The eighth item, an apparatus for... and the associated method of making it..;

All of the above is only an example, each technical solution has its own specific characteristics and the composition and structure of the claims may vary accordingly

LIST OF REFERENCES, PATENT AND LICENCE INFORMATION

ANNEX 1

United States Patent Application

20220200839

Kind Code

A1

Sanger, John

June 23, 2022

SYSTEMS AND METHODS FOR IMPROVING SMART CITY AND SMART REGION ARCHITECTURES

Abstract

Improved systems, methods, and architectures to enhance decision-making
in *Smart* Cities and *Smart* Regions. A system includes an index structure including a first hierarchical data structure including a first hierarchical score based on a plurality of first-level elements, each of the plurality of first-level elements having a respective weighting, and a second hierarchical data structure including a plurality of second hierarchical scores based on a plurality of second-level elements, each of the plurality of second-level elements having a respective weighting, such that the first hierarchical score is based on the plurality of second hierarchical scores through an index factor. and a computer-implemented regional monitor engine to manage local access to a plurality of external data sources to coordinate writes to the index structure.

ANNEX 2

United States Patent Application

20220172208

Kind Code

A1

<u>**Cella. Charles Howard. et al.**</u>

<u>**June 2, 2022**</u>

SYSTEMS AND METHODS FOR CONTROLLING RIGHTS
RELATED TO DIGITAL KNOWLEDGE

Abstract

Systems and methods for controlling rights related to digital
knowledge are disclosed. The system may include an input system to
receive a three-dimensional (3D) printer instructions set for printing
a 3D project. a tokenization system to tokenize the digital
knowledge and a ledger management system to store the tokenized
digital knowledge. The system may further include a *smart* contract
system to implement a *smart* contract via the distributed ledger,
perform a *smart* contract action with respect to the tokenized digital
knowledge in response to an occurrence of a triggering event. The
distributed ledger includes a plurality of cryptographically linked
blocks distributed over a plurality of nodes of a network.

ANNEX 3

United States Patent Application

20220172207

Kind Code

A1

Cella. Charles Howard. et al.

June 2, 2022

COMPUTER-IMPLEMENTED METHODS FOR CONTROLLING RIGHTS RELATED TO DIGITAL KNOWLEDGE

Abstract

A computer-implemented method for controlling rights related to digital knowledge is disclosed. The method includes creating and managing a distributed ledger which includes a plurality of blocks linked via cryptography distributed over a plurality of nodes of a network. The method further includes implementing and managing a *smart* contract which includes a triggering event and a *smart* contract action. The method further includes receiving, tokenizing, and storing an instance of the digital knowledge in the distributed ledger. The method includes managing, rights of control of and access to the tokenized digital knowledge based on the *smart* contract, and performing, in response to an occurrence of the triggering event, the corresponding *smart* contract action with respect to the tokenized digital knowledge.

United States Patent Application

20220148579

Kind Code

A1

Park. Youngchoon. et al.

May 12, 2022

BUILDING SYSTEM WITH AN ENTITY GRAPH STORING SOFTWARE LOGIC

Abstract

One or more non-transitory computer readable media contain program instructions that, when executed, cause one or more processors to: receive first raw data including one or more first data points generated by a first object of a plurality of objects associated with one or more buildings. generate first input timeseries according to the one or more data points. access a database of interconnected *smart* entities, the *smart* entities including object entities representing each of the plurality of objects and data entities representing stored data, the *smart* entities being interconnected by relational objects indicating relationships between the *smart* entities. identify a first object entity representing the first object from a first identifier in the first input timeseries. identify a first data entity from a first relational object indicating a relationship between the first object entity and the first data entity. and store the first input timeseries in the first data entity.

United States Patent Application

20220148356

Kind Code

A1

<u>Sinha. Sudhi R.. et al.</u>

<u>May 12, 2022</u>

BUILDING CAMPUS WITH INTEGRATED SMART ENVIRONMENT

Abstract

A *building* campus with an integrated *smart* environment can provide frictionless access control and device management services among other benefits. A method for providing frictionless access control includes maintaining a directory of individuals associated with the *building,* receiving identity information related to an individual seeking authorization to enter an access point in the *building* from at least one access control device, identifying the individual by comparing the identity information to the directory, identifying a user device associated with the individual using the directory, sending an authorization request to the user device, and authorizing the individual to enter the access point upon completion of the authorization request. A system for providing device management services includes registering devices from different manufacturers to a directory and authorizing one or more users to access and monitor parameters associated with each device.

ANNEX 6

United States Patent Application

20220112766

Kind Code

A1

Snader. Kenneth. et al.

April 14, 2022

VISUAL SECURITY AND ENVIRONMENTAL SELF-ADJUSTING WINDOW

Abstract

A *smart* window including a motorized shade is provided, particularly where the *smart* window includes a frame portion having a first subframe and a second subframe for mounting the *smart* window and routing the motor wirings. In a described embodiment, the *smart* window comprises: a frame portion including a wiring chase positioned internal to the frame portion and a glass portion. The glass portion may comprise a motorized shade, the motorized shade including a motor, a motor wiring, and a shade roll. a first pane of glass attached to the frame portion on an exterior side of the *smart* window. and a second pane of glass attached to the frame portion on an interior side of the *smart* window. wherein the frame portion surrounds the glass portion, wherein the motorized shade is attached to the frame portion between the first pane of glass and the second pane of glass by a hanging system, and wherein the motor wiring is positioned internal to the frame portion and the wiring chase.

ANNEX 7

United States Patent Application

20220093347

Kind Code

A1

<u>**Fadell. Anthony M.. et al.**</u>

<u>**March 24, 2022**</u>

SMART WALL SWITCH CONTROLLER

Abstract

This patent specification relates to various *smart-home* systems. Such a system may include a battery-powered *smart* home device that communicates using a first wireless protocol characterised by relatively low power usage and relatively low data rates. Such a system may further include a *smart* wall outlet device. The *smart* wall outlet device may include wireless communication circuitry comprising a first wireless interface and a second wireless interface. The first wireless interface may be configured to communicate with the battery-powered *smart* home device using the first wireless protocol. The second wireless interface may be configured to serve as a communication bridge between the battery-powered *smart home device* and a wireless network that uses a second communication protocol characterized by relatively higher power usage and relatively higher data rates.

United States Patent Application

20220083032

Kind Code

A1

KIM. Sun Kak. et al.

March 17, 2022

SMART FACTORY MONITORING SYSTEM

Abstract

A *smart* factory monitoring system of the present disclosure includes a sensor module including a plurality of sensors mounted on various facilities of a factory to detect different physical properties and converting communication protocols of a plurality of sensor data into an integrated protocol, an integrated management module integrating and managing data measured by the sensor including the plurality of sensor modules through a pattern or trend analysis, a management server backing up and managing data on the pattern or trend analysed by the integrated management module and providing the data according to a request of an external device, and a manager terminal connected to the integrated management module or the management server to perform monitoring at a remote location. Accordingly, it is possible more easily to detect abnormal signals of equipment through trend analysis or pattern analysis for data measured by each sensor.

ANNEX 9

United States Patent Application

20220070014

Kind Code

A1

Schwartz. John H.. et al.

March 3, 2022

SYSTEM AND METHOD FOR A CONTROL SYSTEM FOR
MANAGING SMART DEVICES IN A MULTIPLE UNIT
PROPERTY ENVIRONMENT

Abstract

A control system for managing *smart* devices in a multi-unit
property environment according to various aspects of the present
technology may comprise a plurality of *smart* home systems,
wherein a single *smart* home system is installed in each unit
throughout the multi-unit property. The system may further
comprise a *smart* community system comprising a plurality of
community controllable *smart* devices installed throughout the
multi-unit property. The *smart* home systems and the *smart*
community system may each be configured to communicate with an
access control system and infrastructure control system to allow
individual residents to control the *smart* home system associated
with their residence and have limited control over at least a portion
of the community controllable *smart* devices through a single user
interface.

United States Patent Application

20220051515

Kind Code

A1

<u>**Schmidt. Mark Christopher. et al.**</u>

<u>**February 17, 2022**</u>

DOUBLE-SIDED STORAGE LOCKER SYSTEMS ACCESSED AND CONTROLLED USING MACHINE-READABLE CODES SCANNED BY MOBILE PHONES AND COMPUTING DEVICES

Abstract

A double-sided ride storage locker system deployed at a park facility with ride sites, including a system integrated with a facility ride management system, for automated management and control over the operation of locker-rental state indication lights displayed as rented on the egress side of the double-sided ride storage locker system. The double-sided ride storage locker system provides guest visitors with access control enabled by scanning multi-level machine-readable codes using mobile scanning computing systems, such as web-enabled smartphones with digital cameras and mobile application support. The storage locker system supports automated modes of discovering and finding where a guest's rented locker is located within the facility and its sites at any moment in time, simply by using the guest's *smart* phone to scan a device-level code, a site-level code, a facility-level code or a discovery-level code posted anywhere within the facility or any site, without need for using a physical locker lookup kiosk or other conventional systems and methods.

United States Patent Application

20220044673

Kind Code

A1

Park. Youngchoon. et al.

February 10, 2022

BUILDING SYSTEM WITH ENTITY GRAPH COMMANDS

Abstract

One or more non-transitory computer readable media contain program instructions that, when executed, cause one or more processors to: receive first raw data including one or more first data points generated by a first object of a plurality of objects associated with one or more buildings. generate first input timeseries according to the one or more data points. access a database of interconnected *smart* entities, the *smart* entities including object entities representing each of the plurality of objects and data entities representing stored data, the *smart* entities being interconnected by relational objects indicating relationships between the *smart* entities. identify a first object entity representing the first object from a first identifier in the first input timeseries. identify a first data entity from a first relational object indicating a relationship between the first object entity and the first data entity. and store the first input timeseries in the first data entity.

ANNEX 12

United States Patent Application

20220004672

Kind Code

A1

Santarone. Michael S.. et al.

January 6, 2022

APPARATUS FOR DISPLAYING INFORMATION ABOUT AN ITEM OF EQUIPMENT IN A DIRECTION OF INTEREST

Abstract

Methods and apparatus for determining information about an item of equipment in a direction of interest based on coordinates derived from wireless communication between wireless transceivers. A *smart* device assembly is operative to communicate via multiple antennas with a reference point transceiver. A set of coordinates is generated indicating a relative position and/or angle of the wireless transceiver in relation to the reference position transceiver. A query may be made based upon the relative position and angle of the wireless transceiver in relation to the reference position transceiver. A response to the query may include a human readable interface indicating one or more of: direction of travel, a virtual image based upon location and location and direction, and annotative and pictorial information.

ANNEX 13

United States Patent Application

20210407229

Kind Code

A1

SCHOENFELDER. Luke A.. et al.

December 30, 2021

SMART *BUILDING INTEGRATION* AND DEVICE HUB

Abstract

Embodiments are generally directed to systems, devices, methods, and techniques to control devices via a mobile device platform in a *smart building* system. Embodiments further include techniques to determine a device of the *smart building* system and an action to perform by the device. The techniques include establishing, a connection with a *smart* lock of the *smart* system, and communicating a request to perform the action to the *smart* lock of the *smart* system.

ANNEX 14

United States Patent Application

20210397166

Kind Code

A1

<u>**Sayyarrodsari. Bijan. et al.**</u>

<u>**December 23, 2021**</u>

INDUSTRIAL AUTOMATION CONTROL PROGRAMME
GENERATION FROM COMPUTER-AIDED DESIGN

Abstract

An industrial control programming development platform simplifies
the generation of an industrial control program and associated tag
definitions by generating at least a portion of the control program
and tag definitions based on analysis of digital engineering drawings
of an automation system to be monitored and controlled. This
drawing-based program generation includes creation and
configuration of *smart* data tags that model and contextualize
controller data at the device level for processing by higher-level
analytic systems. This device-level contextualization can be based in
part on inferences drawn from the digital engineering drawings.

United States Patent Application

20210390222

Kind Code

A1

Wodrich. Michael. et al.

December 16, 2021

METHODS OF COMMUNICATING GEOLOCATED DATA
BASED UPON A SELF-VERIFYING ARRAY OF NODES

Abstract

Methods and apparatus for verifying respective positions of Nodes
based upon ultrawideband wireless communications between nodes
included in an array. Values for variables derived from multiple
wireless transmissions between the nodes are aggregated, and a
position of a particular node may be determined based upon multiple
data sets generated by multiple communications of disparate Nodes.
Data is transmitted to a *smart* device based upon a position of a
particular node and a direction of interest. In addition, the presence
of an obstacle to wireless communication between some nodes may
be derived from the data sets. A user interface may provide a
pictorial view of positions of all or some Nodes in an array, as well
as a perceived obstruction.

Printed by Books on Demand GmbH, Norderstedt / Germany